# Kindergarten Science Volume 2

© 2013 Todd Deluca
OnBoard Academics, Inc
Newburyport, MA 01950

800-596-3175
www.onboardacademics.com

# Table of Contents

# Living vs. Nonliving

Sort these objects.
Write the name of the object in the living or the non living box.

Books

Tree

Dog

Computer

Backpack

Boy

Caterpillar

Pen

Four Characteristics that Living Things Have in Common

Living things have four main things in common. All living things grow. For example humans grow from a baby to an adult.

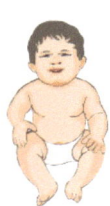

While most plants grow from a seed to a plant.

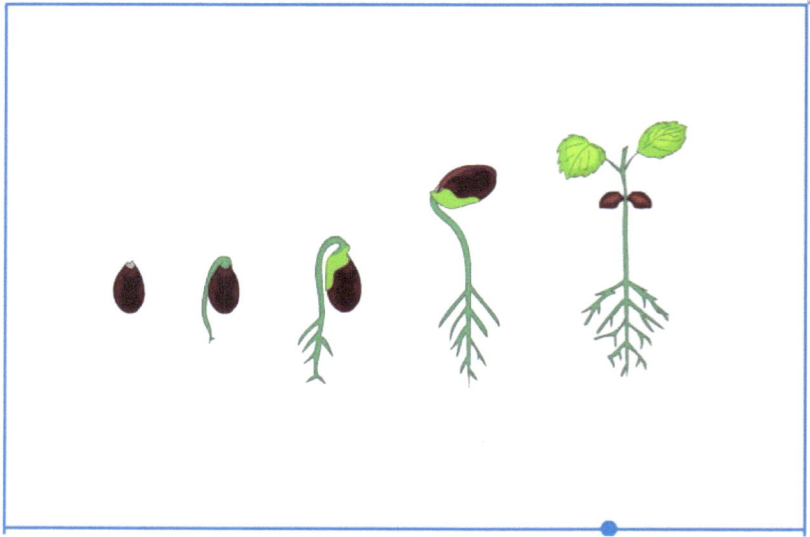

All living things need food and water to survive.   For example animals eat plants or animals or both.

Plants make their own food by using the sun's energy and the nutrients from the soil.

Both plants and animals need water to  survive.

**All living things have four main things in common: they grow, they need food and water, they reproduce, and they react to changes around them.**

All living things reproduce.  This means that they have infants or babies.  Some infants look much like their parents…

…while others look very different.

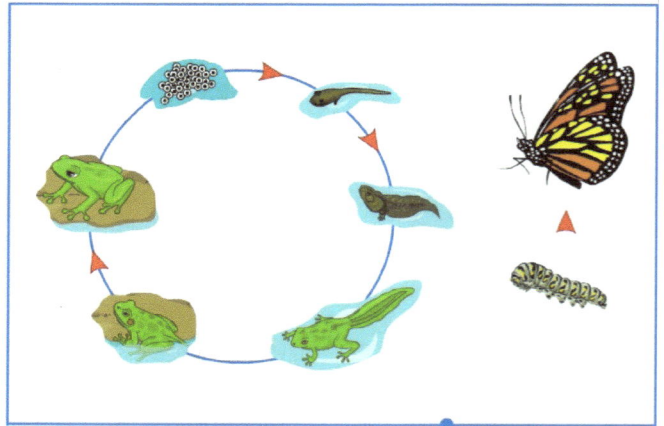

Most plants reproduce from seeds.

All living things react to change. For example the fur on the snowshoe hair change from short and brown to thick and white during the winter.

The roots of a plant will grow toward the source of water.

Which of these things grow?

Label each object.

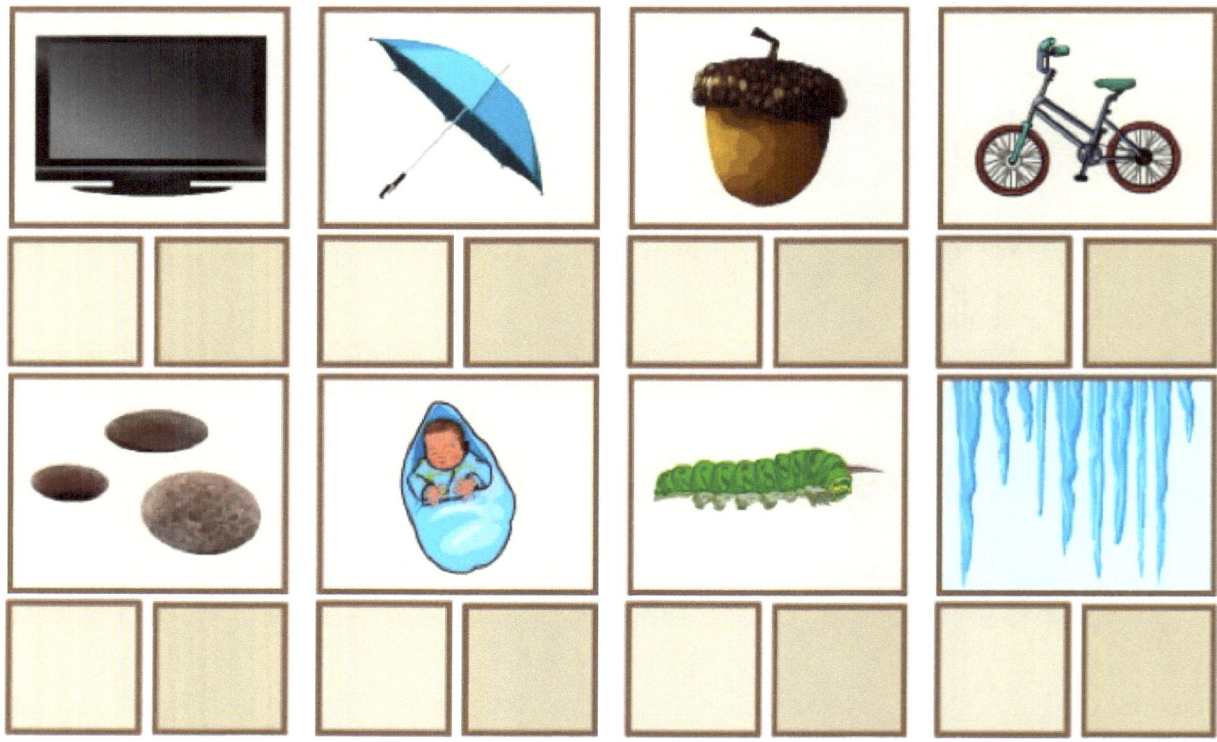

Which of these things needs food and water?
Label each item.

Which of these things react to changes around them?
Label each object.

Which of these things reproduce?
Label each item.

## Living vs. Nonliving Quiz

**All living things _____ over time.**

**All living things need _____ and _____.**

**All living things _____ to changes around them.**

**All living things _____; this means that they have babies.**

water     air     grow     plants     senses

reproduce     react     food     shrink

# All About Air

Full of Air.

Circle the object that is not filled with air.

Although you can't see it, air is all around us. But what is air exactly? Air is an invisible gas that is made up of a mixture of many different elements.

It's mostly made up of two gasses called nitrogen and oxygen. Air also contains small amounts of other gasses such as carbon dioxide. It also contains other elements like tiny dust particles. You might be surprised to learn that air also contains water. Water exists in the air in a gaseous state that we call water vapor. In dry places there is very little water in the air. On a hot sticky day, there is lots of water in the air.

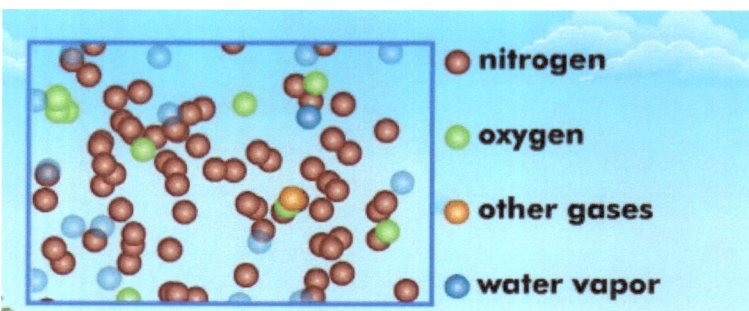

Although air is invisible you can see it when you blow bubbles under water or indirectly when you blow up a balloon on or inflate a bicycle tire.  You can also feel air on a windy day or when you wave you hand around.

The earth is surrounded by a blanket of air that we call the earth's atmosphere and it is essential to life on earth.  That's because when animals and plants breath air they use its ingredients to help them survive and grow.  For example, when we breath in air our bodies use the oxygen in the air.  When plants breath in air they use the carbon dioxide in the air.   Fish don't breath in air but like other animals they do need oxygen which they are able to take from the water.

Which of these plants and animals breath air?
Sort the items by writing their name in the proper box.

| breathes air | doesn't breathe air |
| --- | --- |
| | |

Tree    Boy    Zebra    Tomato Plant    Eagle    Fish    Grasshopper    Orca Whale

Some things that are not living need air too.
Look what happens to these things when we remove the air.

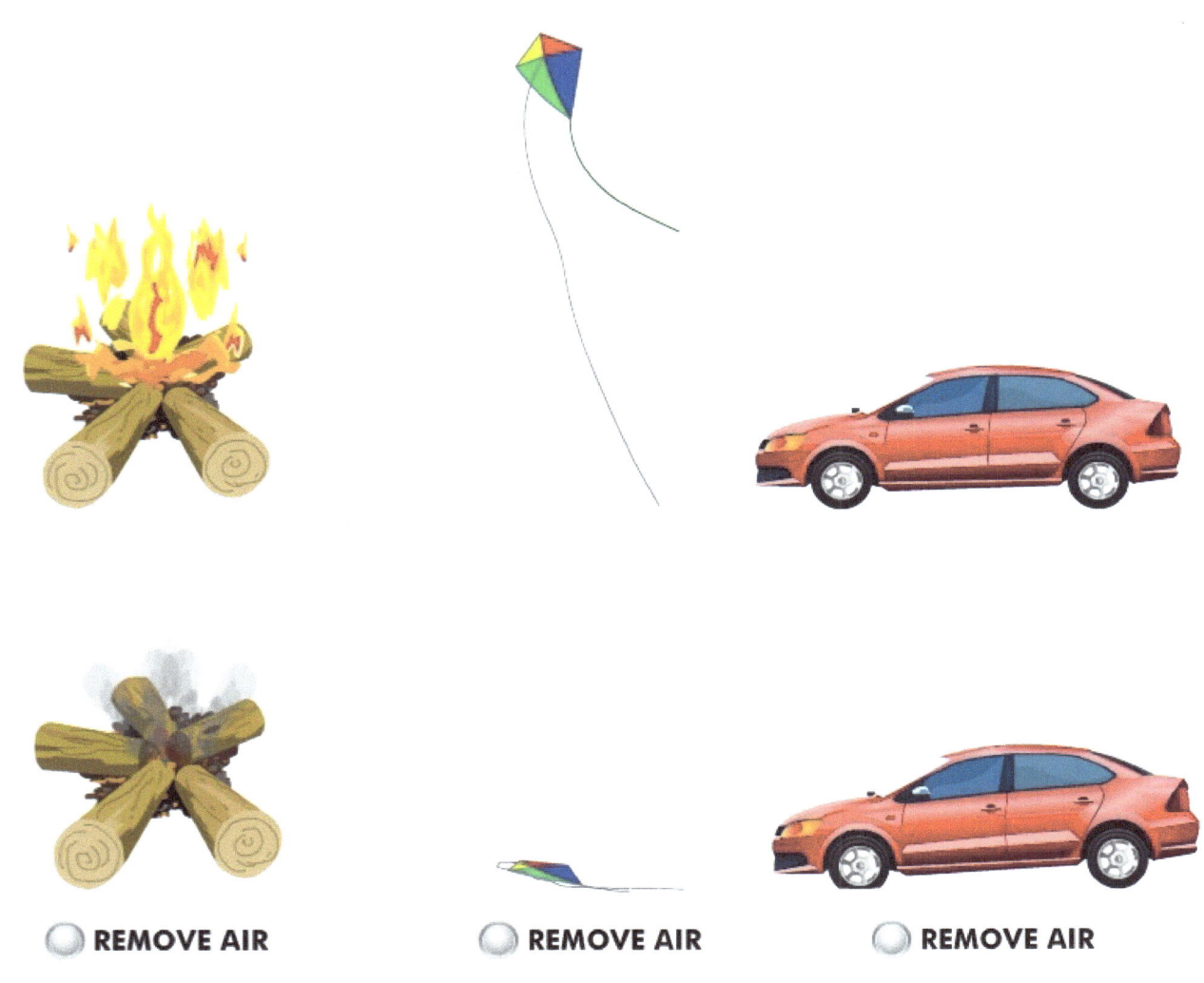

◯ REMOVE AIR    ◯ REMOVE AIR    ◯ REMOVE AIR

Which of the non living things need air to work.
Sort the items in the correct boxes.

| needs air | doesn't need air |
|---|---|
|  |  |

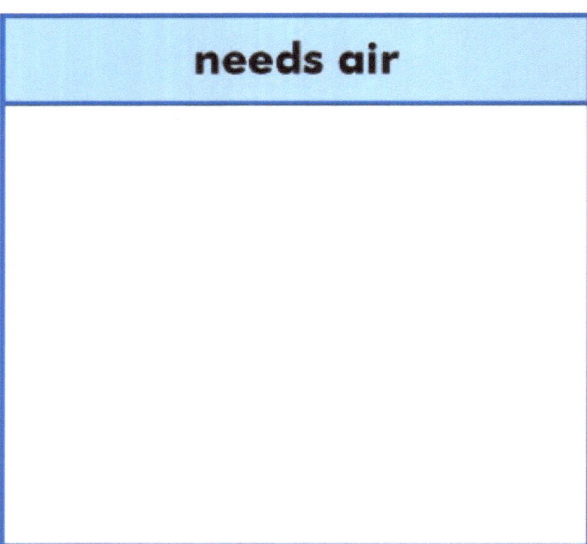

Bat & Ball
Life Vest
Book
Matches

Airplane
Parachute
Skate Board
Scissors

Water in the Air

If you've ever noticed a puddle in the playground that dried up on a warm day you may have wondered where the water went.  The answer is into the air. The sun's rays warm

the water and turn it into a gas which we call water vapor.  The water vapor then becomes part of the air.  We call this process evaporation. We use the term humidity to

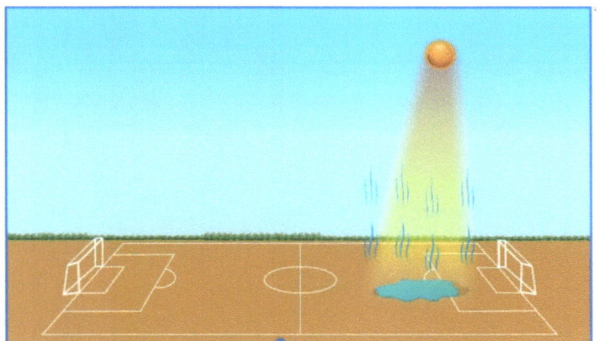

describe the amount of water vapor in the air.  If there is a lot of water in the air then the air feels hot and sticky and we say the humidity is high.

Which of these forms of transportation pollute the air?
Color the cloud in black to signify pollution.

> **To pollute means to make dirty. Air gets polluted when gasoline, wood, and coal are burned as particles from these substances become part of the air.**

# All About Air Quiz.

Answer True or False

Air sinks in water.

☐

If there was no air, birds and kites could not fly.

☐

Fire can't burn unless there is air.

☐

Air can hold water.

☐

Air is mostly made of the chemicals nitrogen and oxygen.

☐

There is lots of air in outer space.

☐

Air is normally invisible.

☐

Electric cars pollute the air.

☐

# Reduce, Reuse and Recycle

Do you know your three Rs?
Label the actions below.

**recycle     reduce     reuse**

**Reducing** the amount of energy that we use, choosing **reusable** products such as reusable shopping bags and water bottles (or finding a new way to use something) and **recycling** as much as we can, all help to conserve Earth's resources and to protect our environment.

The Life of a Soccer Ball

It's important to reduce, reuse and recycle because everything we use and consume has a surprising large impact on our environment.

reduce          reuse          recycle

For example that soccer ball that you are kicking around ant recess is probably made of plastic.

As you may know plastic comes from crude oil that is drilled from the ground or the ocean.  The crude oil is transported to a factor where its turned into plastic.  This plastic is then transported to another factory where the soccer ball is made and where packaging material is added.  The soccer balls are then placed in a large container that is transported to a dock where a large ship transports them to your country.

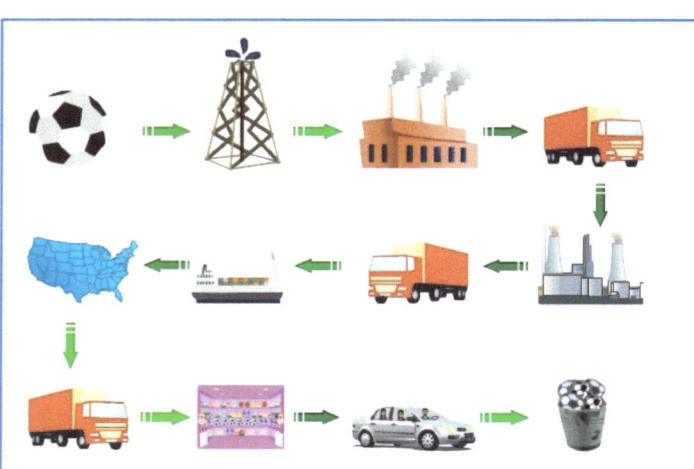

When the ship arrives another truck picks up the container and the soccer balls are delivered to various stores throughout the country.

You drive to the store with your mom an dad to buy your new soccer ball. You have a lot of fun with your soccer ball but in a year or two the ball isn't looking so good so it is taken to the dump.

Plastic doesn't break down so well so the ball may be in the landfill for over a 1,000 years. Or it might be burned in an incinerator that adds pollution to the atmosphere.

As you can see the soccer ball doesn't use any energy directly but a lot of energy is used to make the ball and transport it which causes pollution.  If you don't find another way to use it, it will sit in a landfill long after you are gone.

Can you think of any ways to reuse a soccer ball?

1. _____

_____

_____

2. _____

_____

_____

3. _____

_____

_____

## Ideas

**You could pass it on to a younger sibling, donate it to a neighborhood dog, or use it to mark where your car should stop in the garage. When you're reducing, reusing and recycling, you need to get creative!**

How can you use less energy and resources?
Find a match below and draw it in the correct box.

                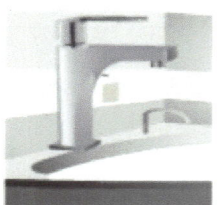

Changing lightbulbs can save energy.

Circle the type of lightbulb that uses the least energy.

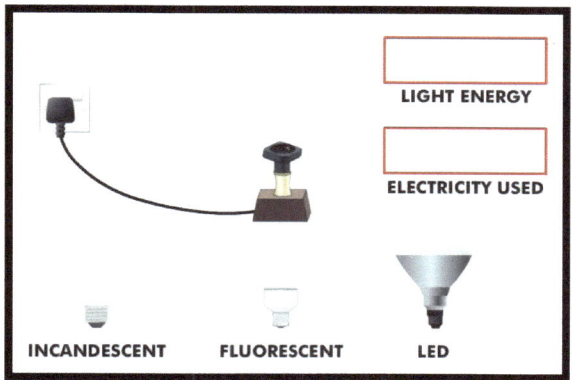

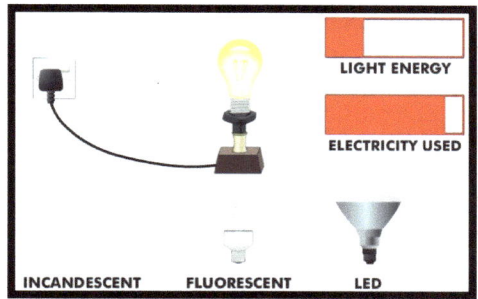

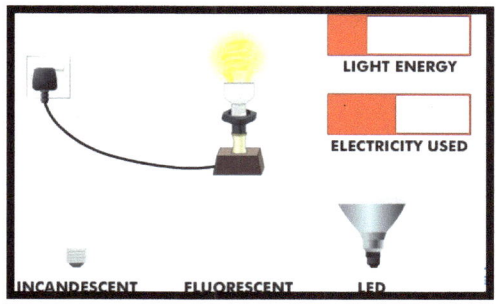

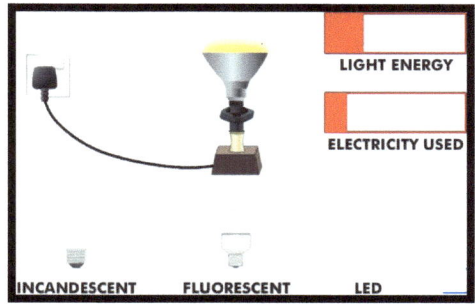

As you can see, switching bulbs can significantly reduce the amount of energy you use at home, and it can also save you a bunch of money.... as much as $30 a month for many households.

Reduce

**Electronic devices use up a lot of energy when they're left on in standby mode. When you connect your electronic devices to a power strip, it's easy to turn them off. Drag the amounts to show how much money you could save each year by doing this.**

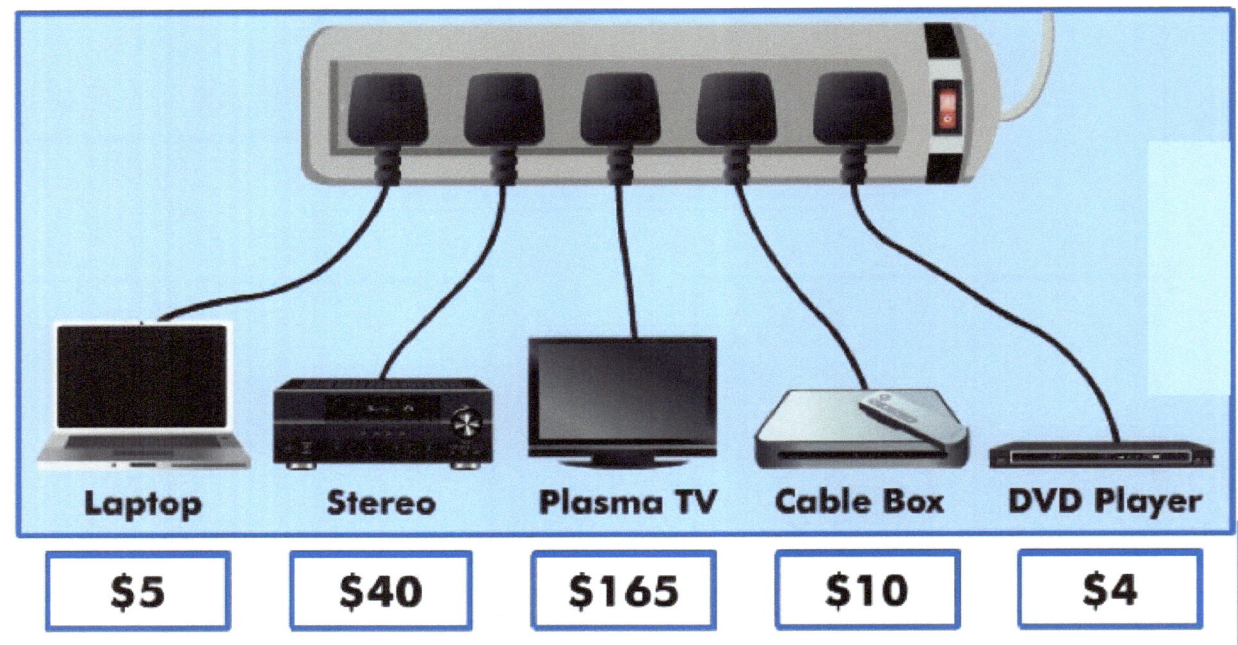

| Laptop | Stereo | Plasma TV | Cable Box | DVD Player |
|--------|--------|-----------|-----------|------------|
| $5 | $40 | $165 | $10 | $4 |

Reuse

Match the reuse tip with the proper illustration by entering the correct letter into the circle

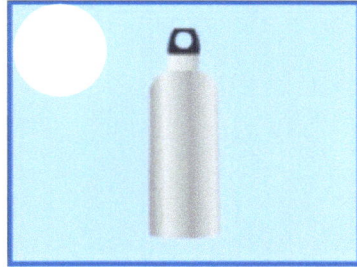

**A** Use a reusable water ball. Disposable water bottles waste energy and pollute the earth.

**B** Start a compost pile. You can compost all of your yard clippings and other yard waste along with all your left over vegetables and fruit. The compost will help to improve the soil in your

**C** When you need to buy something, consider buying a used item from a yard sale or consignment store. When you no longer need your items consider having a yard sale or donating them instead of throwing them into the trash.

**D** Use reusable shopping bags. Did you know that the average person uses over 350 plastic or paper shopping bags a year? If you use reusable shopping bags you are cutting down on a lot of waste.

Recycle

Circle the items that can go in the compost pile.

## Reduce, Reuse and Recycle Quiz

1. Reducing, reusing and recycling helps to conserve the Earth's resources and to protect our environment. True or false?

2. What can be done immediately to conserve water in a leaking tap?
   a. Fix the Tap
   b. Put a bucket under the drip
   c. Do nothing

3. Your school bag does not use electricity or cause pollution, so it doesn't have any impact on the environment. True or false.

4. What are you doing when you replace your light bulb with a fluorescent one?
   a. Reduce
   b. Reuse
   c. Recycle

5. You have grown out of some of your clothes. Is there a way to recycle them?
   a. None. I can only dump them
   b. I can donate them to charity

# Habitats

Where do these animals live?
Write your answer in the box provided.

| Jungle | Desert | Ocean | Lake | Arctic |

**The environment in which an animal lives is called its habitat.**

What key things do an animal need from their habitat to survive?

**Food.** Animals must have food to survive. Some habitats provide food all year round for animals, while other habitats provide food for only part of the year. In this case, animals must change their habitat in order to find food.

**Space** means the amount of living room that an animal enjoys. If there isn't enough living space in a habitat, too many animals are competing for limited food and water resources, while overcrowding can lead to diseases.

**Water.** All animals must have water in order to survive, but how much water an animal needs, and where the water comes from varies quite a lot. Some animals drink from rivers and lakes, while other animals can get the water they need from food.

**Shelter** is an essential part of a habitat, providing animals with protection from the weather and from predators (other animals that might eat them), and a place where they can sleep, and safely bring up their young.

Label this tiger's habitat.

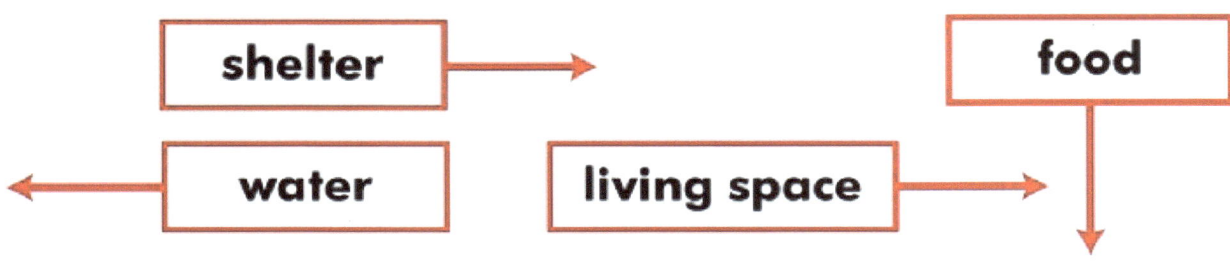

Choose a suitable plant and animal for each habitat.

(you might have to research the plant names)

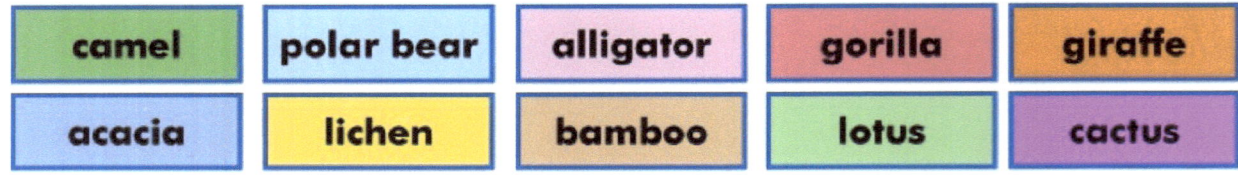

What would help this animal to eat the leaves on the tree?_____

Animals and plants adapt to their environment by developing special characteristics, sometimes over millions of years. These characteristics help them adjust to the conditions of their habitat. For example, a giraffe's long neck enables it to reach leaves on the acacia tree and to spot predators. We call this adaptation.

Connect the animal with the correct description.

My spotted fur helps me to hide in the rain forest. My powerful jaw helps me to kill large animals and even break turtle shells. I am short and stocky and an excellent swimmer

My chiseled shaped teeth are great for cutting down trees. My flat shaped tail and webbed feet are great for getting around rivers and ponds.

My wide flat reed and feathery roots help me to float on the water.

My powerful legs are great for jumping and swimming and I can lasso bugs with my long sticky tongue. I have adapted to cold weather by hibernating which means I sleep until its warm again.

## Habitat Quiz

1. A habitat is the environment in which an animal lives. True or false?

2. Habitats differ for different animals. True or false?

3. The arctic tundra is a type of habitat. True or false?

4. The characteristics that animals develop to adjust to their environment are called _____.
   a. adaptation
   b. survival

5. Polar bears have thick fur that help them survive the cold. True or false?

6. _____ have a hump that sores fat and allows them to survive for a long time without food and water.
   a. Camels
   b. Giraffes
   c. Elephants
   d. Monkeys

# How Plants Make Food

Do you know your plant parts?

Connect the plant part with its name.

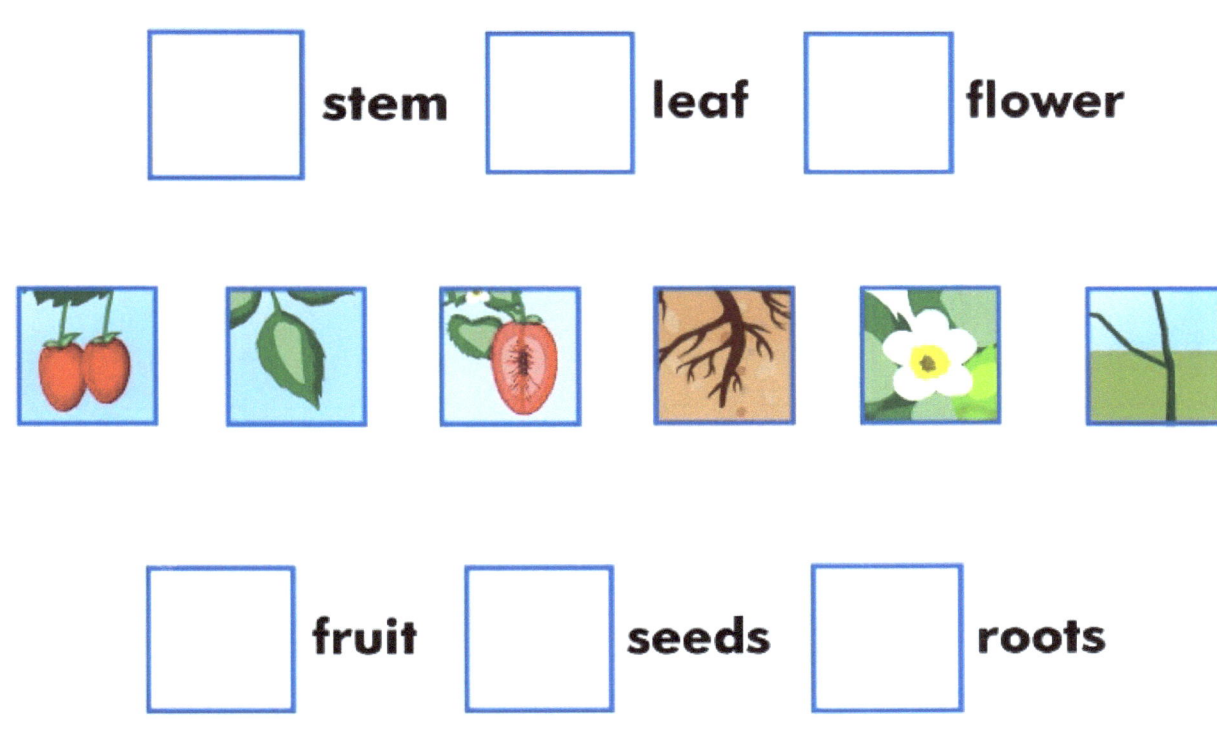

stem    leaf    flower

fruit    seeds    roots

All of the food that we eat comes from plants or from animals that eat plats.  But what do plants eat.

It might look like plants are just there doing nothing but unlike humans and other animals plants are busy making their own food.  They do this using an amazing process called photosynthesis.

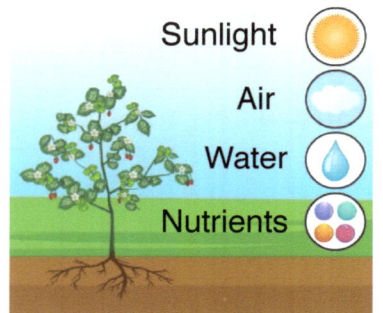

Sunlight
Air
Water
Nutrients

The four ingredients in the process of **photosynthesis** are sunlight, water, air and the nutrients from soil.

The main action happens in the plants leaves. The plant uses tiny opening in its leaves to capture sunlight and air.

Water and soil nutrients enter through the opposite direction through the roots and then through the stems and on to the leaves.  With all the ingredients in place, the leaves turn the sunlight, air, water and nutrients into sugar. This is the plant's food source that will help it to survive and to grow.

Arrange the steps to show how plants make food.

Where do sunlight, air, water and nutrients enter the plant? Draw sunlight, air, nutrients, and water next to the plant part that they use to enter the plant.

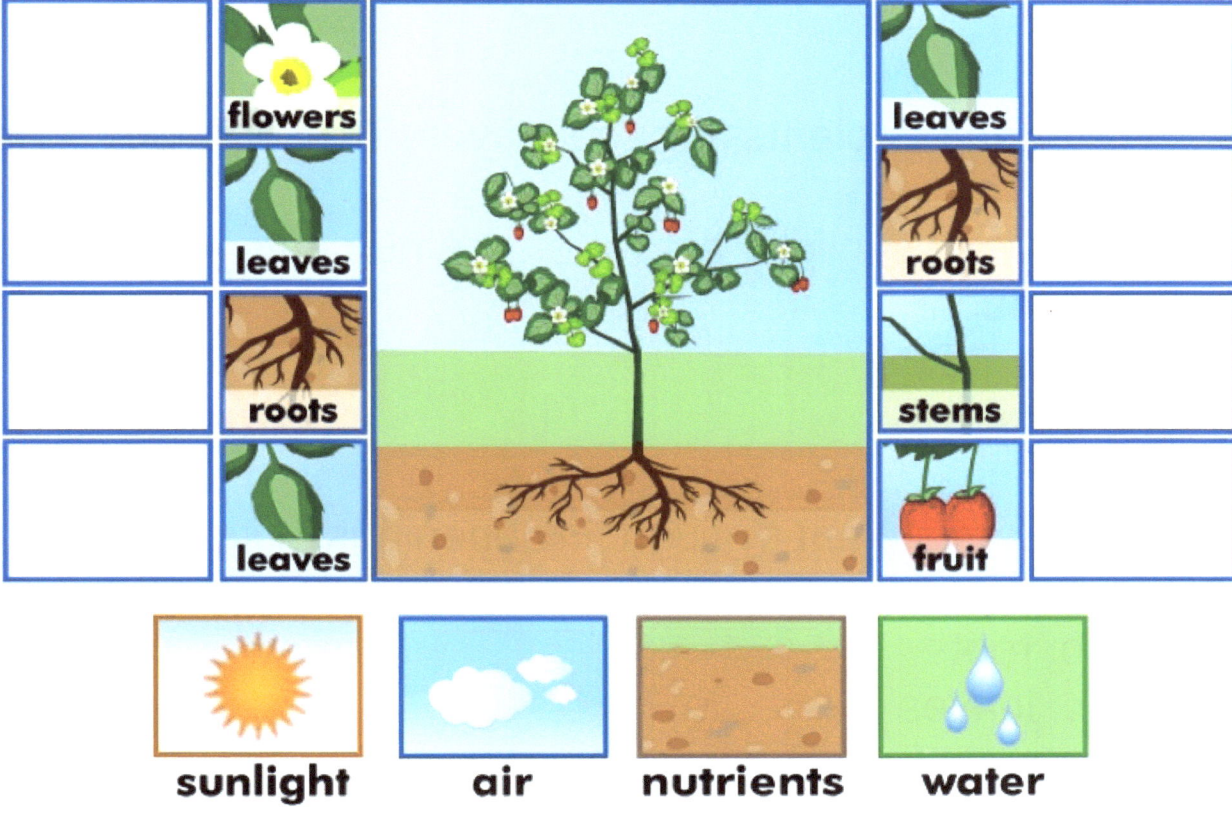

## How Plants Make Food Quiz

1. Plants make their own food.  True or false?

2. The process in which plants make their food is called
   _____.
   - a. photosynthesis
   - b. food synthesis
   - c. Nutrient synthesis

3. The four main ingredients for photosynthesis to take place are sunlight, air, water and nutrients. True or false?

4. Water enters plants through leaves. True or false?

5. Plants turn sunlight, air, water and nutrients into ____.
   - a. sugar
   - b. salt
   - c. fat

6. Air enters a plant through tiny openings in its _____.
   - a. stem
   - b. roots
   - c. leaves

Newburyport, MA 01950

1-800-596-3175

OnBoard Academics employs teachers to make lessons for teachers! We create and publish a wide range of aligned lessons in math, science and ELA for use on most EdTech devices including whiteboard, tablets, computers and pdfs for printing.

All of our lessons are aligned to the common core, the Next Generation Science Standards and all state standards.

If you like our products please visit our website for information on individual lessons, teachers licenses, building licenses, district licenses and subscriptions.

Thank you for using OnBoard Academic products.

www.ingramcontent.com/pod-product-compliance
Lightning Source LLC
Chambersburg PA
CBHW040752200526
45159CB00025B/1860